## VELOCIRAPTOR

Many dinosaurs once roamed Earth.

They lived long before people.

Some dinosaurs became **fossils** after they died.

**Paleontologists** study fossils to learn about dinosaurs.

WHEN DINOSAURS RULED THE EARTH

# THE VELOCIRAPTOR

Written by Tracy Vonder Brink Illustrated by Riley Stark

## TABLE OF CONTENTS



A Crabtree Seedlings Book

Crabtree Publishing
crabtreebooks.com

# School-to-Home Support for Caregivers and Teachers

This book helps children grow by letting them practice reading. Here are a few guiding questions to help the reader with building his or her comprehension skills. Possible answers appear here in red.

## Before Reading:

- What do I think this book is about?
    - *I think this book is about dinosaurs.*
    - *I think this book is about the dinosaur called* Velociraptor.
- What do I want to learn about this topic?
    - *I want to learn how big* Velociraptor *was.*
    - *I want to learn what* Velociraptor *ate.*

## During Reading:

- I wonder why...
    - *I wonder why paleontologists study fossils.*
    - *I wonder how* Velociraptor *used its claws.*
- What have I learned so far?
    - *I have learned that dinosaurs lived before people.*
    - *I have learned that* Velociraptor *lived in Asia.*

## After Reading:

- What details did I learn about this topic?
    - *I have learned that* Velociraptor *had three toes on each foot.*
    - *I have learned that* Velociraptor *ran faster than other dinosaurs.*
- Read the book again and look for the glossary words.
    - *I see the word* ***paleontologists*** *on page 4 and the word* ***carnivore*** *on page 16. The other glossary words are found on page 22.*

*Velociraptor* was a dinosaur that lived about 80 million years ago. *Velociraptor* fossils have been found in Asia.

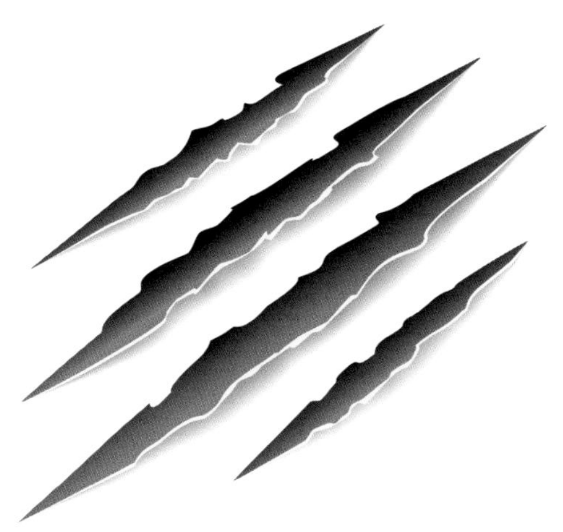

Velociraptor *fossils have been found in China and Mongolia.*

*Velociraptor* was a small dinosaur.

It was about the size of a turkey.

Velociraptor *may have been able to run faster than a human.*

*Velociraptor* walked on two legs.

It ran faster than many other dinosaurs.

*Velociraptor* had three toes on each foot.

One toe on each foot had a sharp claw like a **talon**.

Velociraptor *may have used its claws like hooks to pin down prey.*

*Velociraptor* was a **predator**.
It used its claws to hunt.

*Velociraptor* was a **carnivore**.

It ate small dinosaurs and **mammals**.

Velociraptor *hunted, but it probably also ate animals that were already dead.*

*Velociraptor* was covered in feathers, but it could not fly.

Did *Velociraptor* use its feathers to **attract** mates?

Velociraptor's *feathers might also have helped it stay warm.*

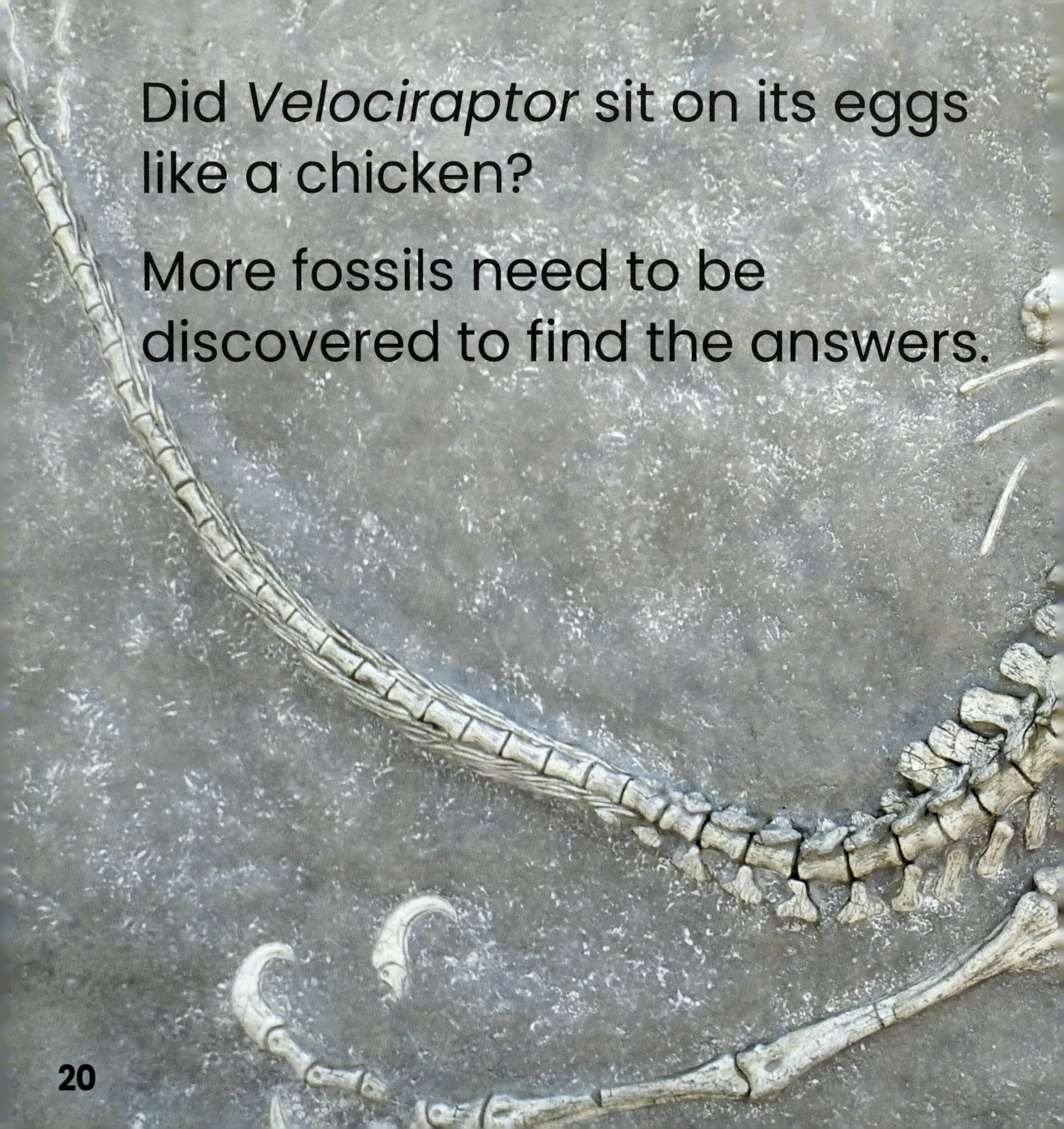

Did *Velociraptor* sit on its eggs like a chicken?

More fossils need to be discovered to find the answers.

# Glossary

**attract** (UH-trakt): To cause interest or attention

**carnivore** (KAR-nih-vor): An animal that only eats meat

**fossil** (FAWS-uhl): The traces, prints, or remains of plants and animals that lived long ago

**mammal** (MAM-uhl): A warm-blooded animal that has hair or fur, and that feeds milk to its young

**paleontologist** (PAY-lee-en-TAW-luh-jist): A scientist who studies fossils to learn about past life on Earth

**predator** (PREH-duh-ter): An animal that hunts and eats other animals

**talon** (TA-luhn): A sharp, curved claw some birds use to catch animals for food

# Index



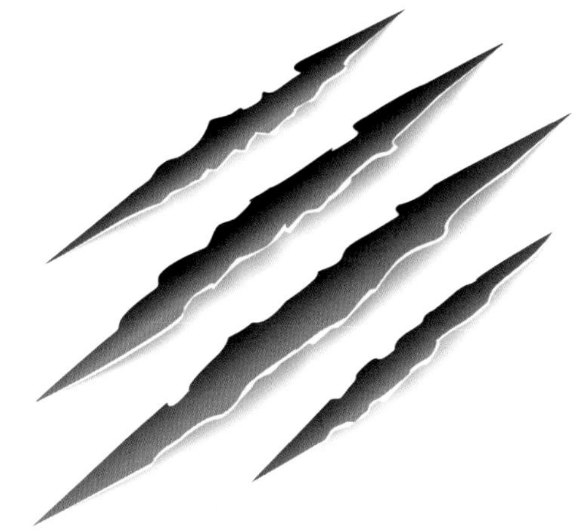

# About the Author

Tracy Vonder Brink loves to learn about science and nature. She has written many nonfiction books for kids and is a contributing editor for three children's science magazines. Tracy lives in Cincinnati, Ohio, with her husband, two daughters, and two rescue dogs. She would like to touch a *Velociraptor* feather.

Written by: Tracy Vonder Brink
Illustrated by: Riley Stark
Designed by: Rhea Wallace
Series Development: James Earley
Proofreader: Melissa Boyce
Educational Consultant: Marie Lemke M.Ed.

Photographs:
Shutterstock: Arpad Benedek: p. 5; Rimma R: p. 7; Noiel: p. 13; Natalia Van D: p. 20-21

**Crabtree Publishing**

crabtreebooks.com 800-387-7650



Printed in the U.S.A./072023/CG20230214

Published in Canada
Crabtree Publishing
616 Welland Ave.
St. Catharines, Ontario
L2M 5V6

Published in the United States
Crabtree Publishing
347 Fifth Ave
Suite 1402-145
New York, NY 10016

**Library and Archives Canada Cataloguing in Publication**
Available at Library and Archives Canada

**Library of Congress Cataloging-in-Publication Data**
Available at the Library of Congress

Hardcover: 978-1-0396-9648-8
Paperback: 978-1-0396-9755-3
Ebook (pdf): 978-1-0396-9969-4
Epub: 978-1-0396-9862-8